BEGINNERS GUIDE TO HYDROPONICS FOR PROFIT

I0429781

Revolutionize Your Harvest: Mastering the Art of Hydroponics through Proven Techniques, Innovative Systems, and Expert Insights to Cultivate Abundant Yields

Lucille Sarron

Table Of Contents

CHAPTER 1 6

  Hydroponics 6

    Overview Of Hydroponics: 6

    Benefits Of Hydroponic Gardening: 7

    Goals And Objectives Of The Book: 8

CHAPTER 2 12

  Knowing How To Use Hydroponics 12

    Definition And Fundamental Ideas: 12

    Background History: 14

    Hydroponic System Types: 14

CHAPTER 3 18

  Advantages Of Gardening In Hydroponics 18

    1. Enhanced Crop Yield: 18

    2. Faster Growth Rates: 19

    3. Water Efficiency: 19

    4. Space Savings: 20

    5. Control Over Growing Conditions: 20

CHAPTER 4 22

  Configuring A Hydroponic System 22

    1. Choosing An Appropriate Location: 22

2. Selecting Your Ideal Hydroponic System: ..23

3. Crucial Instruments And Equipment: 23

4. Comprehending Ph Levels And Nutrients: .24

CHAPTER 5 26

Choosing Plants For Hydroponic Growing.........26

Plant Selection For Hydroponic Growing: 26

The Best Hydroponic Plants: 29

Taking Into Account Crop Selection: 30

Getting Used To Various Hydroponic Systems:
....31

CHAPTER 6 34

Hydroponic Nutrient Solutions And Management
....34

Vital Elements For Plants 34

Blending   And   Organizing   Nutrient
Combinations 36

Resolving Nutrient-Related Problems 38

CHAPTER 7 40

Hydroponic Lighting 40

Light Is Essential For Plant Growth 40

Grow Light Types 42

Duration And Intensity Of Light 43

CHAPTER 8 46

Environmental Management 46

Controlling Temperature: 46

Controlling Humidity: 47

Movement Of Air: 48

CO2 Improvement: 49

CHAPTER 9 52

Management Of Diseases And Pests 52

Typical Hydroponic Pests: 52

Prevention And Control Of Diseases: 54

Management Of Organic Pests: 55

CHAPTER 10 58

Crop Rotation And Harvesting In Hydroponic Systems 58

1. Signs Of Ready For Harvest: 58

2. Harvesting Methods: 59

3. Crop Rotation In Hydroponics: 60

CHAPTER 11 64

Promoting Your Hydroponic Vegetables 64

1. Choosing The Market You Want To Enter:64

2. Establishing A Personal Brand: 65

3. Channels For Sales And Distribution: 67

4. Internet Marketing Techniques: 67

Summary 69

Summary Of Important Takeaways: 70

Motivation For Upcoming Achievements: 71

Closing Remarks And Recognitions: 72

THE END 74

CHAPTER 1

Hydroponics

Greetings from the fascinating realm of hydroponics! This book is your guide to maximizing the benefits of hydroponic farming, regardless of your level of gardening experience. We'll give a quick rundown of the idea of hydroponics in this introduction, go over its many benefits, and lay out the objectives of this book.

Overview Of Hydroponics: Hydroponics is a soilless gardening technique that uses nutrient-rich water solutions to directly supply vital minerals to plant roots. With the help of this cutting-edge gardening technique, you can precisely manage pH levels, nutrients, and other important elements that affect plant growth in a more controlled environment. Because of its effectiveness, conservation of resources, and capacity to yield large amounts of food in a small amount of space, hydroponics has grown in popularity.

Benefits Of Hydroponic Gardening:

Hydroponic gardening has a number of benefits that make it a desirable choice for both commercial growers and hobbyists. Among the principal advantages are:

1. Enhanced Growth Rates: Because hydroponic systems provide direct access to nutrients and ideal growing conditions, plants there frequently develop more quickly than those in conventional soil-based gardens.

2. Water Efficiency: Hydroponic farming is a sustainable option in areas where water is scarce because it requires a lot less water than traditional soil gardening.

3. Maximizing Utilization of Available Space: Hydroponic systems can be arranged vertically or in small areas, allowing for year-round growth.

4. Accurate Nutrient Management: Producers possess exact control over nutrient concentrations, guaranteeing that plants have the proper ratio of vital components for strong development.

5. Decreased Pests and illnesses: Because hydroponic systems do not require soil, there is a lower chance of soil-borne pests and illnesses, which benefits healthier plants.

Goals And Objectives Of The Book:

This book is meant to be your all-in-one resource for everything related to hydroponic gardening. We hope to offer insightful information and useful expertise, regardless of your level of experience with hydroponics—whether you're a novice trying to put up your first system or an expert grower searching for advanced ideas. Among our objectives are:

1. Establish a strong foundation in education by learning about the fundamentals of hydroponics, such as plant physiology, nutrient delivery science, and system design.

2. Practical Advice: Provide detailed instructions for assembling different kinds of hydroponic systems, choosing appropriate plants, and resolving typical problems.

3. Optimizing Outcomes: Offer pointers and techniques to raise your hydroponic garden's output and efficiency to the maximum extent attainable, guaranteeing the greatest outcomes.

4. Inspiration: Showcase the adaptability of hydroponics by examining cutting-edge methods and authentic success tales. This will encourage you to try new things and push the limits of your own hydroponic ventures.

We hope that after reading this book, you will be equipped with the knowledge you need to start a successful hydroponic farming journey as well as the confidence to experiment and tweak your setups in order to produce a bountiful and long-lasting harvest. Together, we will explore the intriguing realm of hydroponics and create a garden that surpasses conventional limitations!

CHAPTER 2

Knowing How To Use Hydroponics

In hydroponics, vital minerals are delivered straight to the roots of the plants via a nutrient-rich water solution, eliminating the need for soil in plant growth. With the exact control it provides over environmental parameters, this soilless farming technology generates higher crop yields, faster growth rates, and more effective use of available resources. The idea of hydroponics is to maximize plant growth by giving them all the nutrients, water, and oxygen they require in a regulated environment.

Definition And Fundamental Ideas:

Compared to conventional soil-based culture, hydroponics provides for improved nutrient absorption and faster growth by supplying plants with critical nutrients through a water-based solution. Among the fundamental ideas of hydroponics are:

1. Nutrient Control: Growers may precisely regulate the solution's nutrient composition in hydroponic systems, adjusting it to the unique requirements of the plants at various phases of growth.

2. Water Efficiency: Because hydroponics uses closed systems with recirculated water, it generally uses less water than traditional soil farming.

3. Oxygenation: Hydroponic systems guarantee that roots receive enough oxygen, which is necessary for plant respiration and healthier plant growth.

4. Management of pH and EC: For maximum nutrient absorption, the pH and electrical conductivity (EC) of the nutrient solution must be kept at the proper values. Growers may closely monitor and modify these parameters with hydroponic systems.

Background History:

The Hanging Gardens of Babylon, one of the Seven Wonders of the Ancient World, is one of the first examples of the hydroponics concept, which stretches back to ancient civilizations. However, with the introduction of nutrient solutions and controlled environment agriculture in the middle of the 20th century, modern hydroponics started to take shape. Different hydroponic systems were investigated by horticulturists and researchers in an effort to increase agricultural yields and resource efficiency.

Hydroponic System Types:

1. Nutrient Film Technique (NFT): In NFT systems, a continuous, shallow stream of nutrient-rich water ishes over the roots. This technique gives the roots access to oxygen while guaranteeing a steady supply of nutrients.

2. Deep Water Culture (DWC): DWC systems use air stones or diffusers to provide aeration while suspending plant roots in a nutrient solution. Effective nutrient uptake is encouraged by the roots being immersed in the nutrient solution.

3. Aeroponics: To achieve optimal oxygenation, aeroponic systems spray nutrient solutions directly onto the suspended roots. This technique is renowned for its quick plant development and effective nutrient delivery.

4. Drip Systems: Using drip emitters, drip systems supply nutritional solutions to the roots of plants. This approach is popular and adaptable to different hydroponic system sizes.

5. Wick Systems: A wick moves nutrient solution from a reservoir to the growing media in wick systems, offering a straightforward and passive way to provide nutrients. It might be less effective for

larger-scale activities, even when it works well for smaller setups.

Gaining knowledge about the numerous kinds of hydroponic systems and the concepts and principles of hydroponics enables farmers to select the best technique for their particular requirements and get ideal plant development in a regulated setting.

CHAPTER 3

Advantages Of Gardening In Hydroponics

Because of its many advantages, soilless gardening has become more and more popular with hydroponic gardening. The following are the main benefits of hydroponic gardening:

1. Enhanced Crop Yield: Plants grown in hydroponic systems have direct access to vital nutrients, which speeds up growth and boosts yield. Hydroponics provides nutrients straight to the roots of plants, saving energy that would otherwise be spent by plants searching for them in traditional soil-based approaches. Increased total productivity from this enhanced nutrient uptake allows for larger crop yields in less time.

2. Faster Growth Rates: Optimal circumstances, including temperature, pH levels, and fertilizer content, are made possible by the controlled environment of hydroponic systems. Plants can develop more quickly if these factors are carefully regulated. With hydroponic gardening, plants may concentrate on growth and development above ground because they no longer have to use energy for significant root development. Because of this, crops produced hydroponically frequently reach maturity earlier than those grown in soil.

3. Water Efficiency: In comparison to conventional soil-based techniques, hydroponic systems are engineered to be extremely water-efficient. These systems usually need less water overall since they use recirculating water. Furthermore, water is managed and given directly to the roots of the plants, lowering the possibility of water waste from evaporation or runoff. Hydroponic systems' capacity to recycle and reuse water helps make agriculture a more ecologically friendly and sustainable practice.

4. Space Savings: Because hydroponic gardening makes optimal use of available space, it is a great option for indoor or urban farming, where available space is frequently at a premium. Plants don't need large soil beds to thrive in a more compact space because they absorb nutrients straight from the soil. This function saves space and is especially useful for vertical farming, which allows crops to be grown in numerous layers in a stacked and regulated manner.

5. Control Over Growing Conditions:

The exact control over growing conditions that hydroponic farming offers is one of its main benefits. It is possible to closely monitor and modify variables including nutrient content, pH levels, and environmental elements like humidity and temperature. This degree of management reduces the effects of outside variables, pests, and illnesses, resulting in healthier plants and a crop that is more consistent. Furthermore, year-round growing is made possible by this control, which is not dependent on seasonal variations outside.

In summary, hydroponic gardening provides a number of advantages that support greater crop production productivity, sustainability, and efficiency. Hydroponics offers a practical and cutting-edge substitute for conventional soil-based farming techniques, whether it is used in large-scale commercial farming or on a smaller scale for personal use.

CHAPTER 4

Configuring A Hydroponic System

Using nutrient-rich water solutions, hydroponics is a technique for growing plants without the need for soil. To achieve ideal plant growth, setting up a hydroponic system requires following several important stages. This guide will assist you in setting up your hydroponic system:

1. Choosing An Appropriate Location:

• Pick a location with lots of natural light, or think about adding more artificial lighting. Make sure it's convenient to monitor and maintain the location.

• Most plants thrive in temperatures between 65 and 75°F (18 and 24°C), so choose a spot with reliable temperature control.

• Sufficient ventilation is essential to keep humidity levels low and to supply new air to support plant growth.

2. Selecting Your Ideal Hydroponic System:

• There are numerous varieties of hydroponic systems, such as aeroponics, deep water culture (DWC), nutrient film technology (NFT), and others. Choose a system according to your preferred level of involvement, available space, and spending capacity.

• While seasoned growers may choose more sophisticated systems like aeroponics, novices may find nutrient film technique (NFT) or deep water culture (DWC) systems easier to operate.

3. Crucial Instruments And Equipment:

• Reservoir: A nutritional solution-holding container.

• Submersible Pump: Makes it easier for the nutrient solution to circulate.

The nutritional solution is delivered to the plants via tubing, pipes, or other conduits.

• Growing Medium: To support plant roots, utilize growing media such as perlite, coconut coir, or Rockwool, depending on the system.

• PH Testing Kit: To guarantee ideal nutrient absorption, periodically check and modify the pH levels of the nutrient solution.

• EC/TDS Meter: To determine the concentration of nutrients in the solution, measure the electrical conductivity (EC) or total dissolved solids (TDS).

• Timer: For constant plant care, set the lighting and nutrient delivery cycles.

4. Comprehending Ph Levels And Nutrients:

• Nutrient solutions are the primary source of critical elements for plant growth in hydroponic systems. Typically, nitrogen, phosphorous, potassium, and trace elements are found in nutrient solutions.

• Keep an eye on pH levels and make adjustments to keep them within the ideal range for nutrient absorption (most plants typically need a pH between 5. 5 and 6.5).

• Restock and tweak the nutrient solution on a regular basis to avoid excesses or deficiencies that could harm the health of the plants.

Always remember to start small and work your way up as you gain hydroponics knowledge. Maintain a close eye on your system and make any adjustments to ensure that your plants are growing healthily. You can experiment with various systems and methods as you gain more experience with hydroponic farming to tailor your setup for particular crops.

CHAPTER 5

Choosing Plants For Hydroponic Growing

Plant Selection For Hydroponic Growing:

The soilless cultivation technique known as hydroponics provides a regulated growing environment for plants. A number of considerations are made while selecting plants for hydroponic farming in order to guarantee ideal development and productivity. Here are some crucial things to remember:

1. High Yield and Quick Growth:

• Select plants with a high potential yield and a comparatively quick growth cycle. This makes it possible to use the hydroponic system more effectively and to change crops more quickly.

2. Needs for Nutrients:

• Take into account the plants' nutrient requirements. Certain crops grow better in hydroponic systems because they can flourish in nutrient-rich water solutions. Certain fruiting vegetables, leafy greens, and herbs work well in hydroponic systems.

3. Efficiency of Space:

• It is possible to construct hydroponic systems to maximize vertical space. Select plants that can be grown in tight spaces or vertically, including dwarf types or cultivars that tolerate pruning well.

4. EC Tolerance and pH:

• The pH and electrical conductivity (EC) tolerances of different plants vary. To prevent nutrient shortages or toxicities, choose plants that will flourish in the pH and EC ranges that your hydroponic system can achieve.

5. Requirements for temperature and humidity:

• Take your hydroponic setup's environmental parameters into account. Select plants that can survive in the ranges of temperatures and humidity that your system can sustain. This is especially crucial for crops with temperature needs for fruiting, such as tomatoes and peppers.

6. Resistance to Disease:

• Choose plant cultivars that are recognized for their ability to withstand common illnesses and pests. Even though hydroponic systems can help with some insect problems, potential disease problems still need to be taken into account.

7. Flexibility in Hydroponic Environments:

• Certain plants are more adapted to hydroponic farming by nature. Because they can adapt to nutrient-rich water solutions, leafy greens (lettuce, spinach, kale), herbs (basil, cilantro), and some fruiting vegetables (tomatoes, peppers) are popular options.

The Best Hydroponic Plants:

1. Leafy greens and lettuce:

• Hydroponic systems are ideal for growing leafy greens like lettuce, spinach, kale, and others. They can be harvested more than once and have comparatively short growing cycles.

2. Herbs:

• Hydroponics is a good fit for herbs like basil, cilantro, mint, and others. They are in demand for a variety of culinary uses and can have a high market value.

3. Tomatoes:

• Hydroponic tomato gardening is very popular, especially with vine or cherry kinds. Although they demand more nutrients, they can produce exceptional harvests.

4. Chilies:

• Hydroponic systems can be used to cultivate spicy and bell peppers with success. Make sure the plants have enough support while they bear fruit.

5. berries:

• Strawberries grow well in some hydroponic systems, such as nutrient film technology (NFT). They need the pH and nutrition levels to be closely monitored.

Taking Into Account Crop Selection:

1. Demand in the Market:

• Take into account the local market's need for particular crops. To ensure a lucrative endeavor, select plants with strong market value.

2. Individual Preferences:

• If you're a home gardener or hobbyist, think about choosing plants according to your culinary requirements and personal tastes.

3. System Dimensions and Architecture:

• The kinds of plants you can grow depend on the dimensions and configuration of your hydroponic system. Make sure the crops you have selected fit the size and setup of your system.

4. Proficiency Level:

• It could be advantageous for novices to begin with crops that are simple to raise before progressing to more difficult types. This can foster experience and confidence growth.

Getting Used To Various Hydroponic Systems:

1. NFT, or Nutrient Film Technique:

• Crops like lettuce and herbs that have shallow root systems work well with NFT systems. The thin layer of nutrient-rich water is ideal for the growth of these plants.

2. Culture in Deep Water (DWC):

• DWC systems work best with plants whose roots can withstand being submerged. Herbs and leafy greens work well in DWC arrangements.

3. Drip Frameworks:

• A large variety of crops can be fitted to drip systems due to their versatility. Depending on the unique requirements of the chosen plants, modify the frequency and length of fertilizer supply.

4. Aeroponics:

• Plants that can survive in an atmosphere with high oxygen levels are suited for aeroponic systems, which mist nutritional solutions onto plant roots. Aeroponic systems are commonly effective when growing herbs and leafy greens.

5. Systems Based on Media:

• Plants that require physical support for their fruit, such as tomatoes and peppers, can be grown in media-based systems with the help of materials like perlite or coconut coir.

To sum up, careful consideration must go into choosing plants for hydroponic growing, taking into account factors like growth traits, nutrient needs, and system compatibility. Growers can maximize yields, optimize resource consumption, and have a good hydroponic farming experience by taking these variables into consideration.

CHAPTER 6

Hydroponic Nutrient Solutions And Management

With hydroponics, vital nutrients are delivered straight to the roots of plants, negating the need for soil in plant growth. Hydroponic systems require efficient nutrient solution management to be successful. In this article, we'll explore the fundamental ideas behind nutrient solutions, such as the nutrients that plants require, how to combine and maintain nutrient solutions, and how to solve problems relating to nutrients.

Vital Elements For Plants

For plants to flourish, they need a variety of vital nutrients. Both macronutrients and micronutrients can be used to broadly classify these nutrients:

1. The macronutrients

• Nitrogen (N)

- Phosphorus (P)

- Potassium (K)

- Calcium (Ca)

Magnesium (Mg)

- Sulfur

2. Small-scale nutrients:

- Iron (Fe)

- Manganese (Mn)

- Zinc (Zn)

- Copper (Cu)

- Lead (Lead)

- Boron

- Chlorine (Cl)

Comprehending the distinct requirements of every plant species and their developmental phases is crucial in formulating a nutrient solution that is harmoniously balanced.

Blending And Organizing Nutrient Combinations

To address the nutritional needs of the plant, the components must be precisely measured and mixed to create a nutrient solution. Take into account these procedures when combining and overseeing nutritional solutions:

1. Water Purity:

• Begin with a clean, contaminant-free source of water. Water should be tested and adjusted to the ideal pH range (often between 5. 5 and 6.5) for hydroponic plants.

2. Formulation of Nutrients:

• To ascertain the proper macronutrient and micronutrient ratios for your particular crop, refer to the manufacturer's recommendations or make use of a reliable nutrition calculator.

3. Observing and Modifying:

• Continually check the electrical conductivity (EC) and pH of the nutrition solution. To make sure that nutrients are available and absorbed, adjust these values as necessary.

4. Feeding Timetable:

• Create a feeding schedule that takes the plant's growth stage into account. The need for nutrients can change between the vegetative and blooming stages.

5. Cleanliness:

• Keep the hydroponic system clean to avoid the accumulation of salts, bacteria, and algae. Flush the system frequently to prevent nutritional imbalances.

Resolving Nutrient-Related Problems

Preventing plant stress and guaranteeing optimal growth requires identifying and treating nutrient-related problems. Typical problems consist of:

1. Inadequacies in nutrients:

• Identify indications such as odd leaf coloring, reduced development, or yellowing leaves (chlorosis). By changing the solution's nutritional concentrations, deficiencies can be addressed.

2. Toxicity of Nutrients:

Burnt leaf tips, curling, or discoloration are possible symptoms. To address toxicity concerns, lower nutrient amounts and make sure that the right flushing is done.

3. Unbalanced pH:

• ApH imbalances can impact the availability of nutrients. To keep the pH within the ideal range for nutrient absorption, test and adjust the pH frequently.

4. EC Variations:

• Keep an eye on EC levels to avoid over- or under fertilization. Adapt the concentrations of nutrients accordingly.

5. Problems with Watering:

• Use appropriate irrigation techniques to prevent under watering or overwatering. This keeps the absorption of nutrients and the intake of oxygen in check.

Hydroponic gardeners may maximize plant growth and generate high yields in their controlled surroundings by knowing the vital nutrients, controlling nutrient solutions correctly, and resolving problems as soon as they arise. Hydroponic nutrition management requires constant observation and modification to be successful.

CHAPTER 7

Hydroponic Lighting

A soilless gardening technique called hydroponics uses nutrient-rich water to promote plant development. For hydroponic systems to be effective, lighting is just as important as fertilizers. Proper photosynthesis, which is the cornerstone of plant development, depends on adequate light. We'll discuss the value of light in hydroponics in this context, as well as the many kinds of grow lights that are frequently employed and the implications of light duration and intensity.

Light Is Essential For Plant Growth

In hydroponic systems, light is a key element that affects the growth and development of plants. Light availability and quality play a critical role in photosynthesis, the process by which plants transform light energy into chemical energy to support their growth. Key components of light in hydroponics are as follows:

1. Photosynthesis: Chlorophyll, the pigment that absorbs light energy, needs light to activate. After that, this energy is used to transform carbon dioxide and water into glucose and oxygen, which are the building blocks needed for plant growth.

2. Light Affects Nutrient Uptake: Plants absorb nutrients in response to light. It promotes the uptake of vital nutrients, resulting in stronger and healthier plant development.

3. Plant Morphology: Light has an impact on a variety of plant morphological characteristics, including general structure, stem length, and leaf size. Plant growth is robust and compacted when exposed to the right amount of light.

Grow Light Types

Selecting the appropriate grow lights is essential for hydroponic achievement. There are several popular varieties of grow lights, each having pros and cons:

1. Fluorescent lights are inexpensive, low-heat producing, and energy-efficient. They are ideal for seedlings and young plants. They might not have the intensity needed for the flowering or fruiting stages, though.

2. High-Intensity Discharge (HID) Lights: HID lights, which include Metal Halide (MH) and High-Pressure Sodium (HPS) lights, produce bright light that is appropriate for all stages of plant growth. While HPS lights work better throughout the flowering and fruiting stages, MH lights are best for vegetative development.

3. Light-Emitting Diodes (LEDs): LED lights may be adjusted to produce precise wavelengths required

for various growth phases. They also produce minimal heat and are energy-efficient. They are adaptable and gaining popularity in hydroponics.

4. Induction lights: Distinguished by their extended lifespan and energy economy, induction lights offer a well-balanced spectrum appropriate for different phases of growth.

Duration And Intensity Of Light

1. Light Duration (Photoperiod): The length of time a plant is exposed to light is essential for regulating its vegetative and reproductive phases. For optimal growth, different plants require different photoperiods. A flowering photoperiod can last anywhere from 12 to 14 hours, whereas a vegetative photoperiod is typically about 18 hours.

2. Light Intensity: The photosynthetic photon flux density (PPFD) is a unit of measurement for light intensity. The needs for light intensity vary amongst

plants. In order to avoid problems like stretching or light burn, it is essential to monitor and modify light intensity according to the growth stage.

In conclusion, boosting plant growth and attaining ideal yields in hydroponic systems require an awareness of the significance of light in hydroponics as well as the selection of the proper grow lights with the suitable length and intensity. A good and effective hydroponic growing experience will result from giving careful thought to these elements.

CHAPTER 8

Environmental Management

Hydroponics Environmental Control: Air Circulation, Temperature, Humidity, and CO2 Enrichment

As a soilless cultivation technique, hydroponics significantly depends on accurate environmental management to guarantee ideal plant growth and development. Hydroponic systems require a precise set of environmental factors, including temperature, humidity, air circulation, and carbon dioxide (CO_2) levels. Let's explore each of these facets individually:

Controlling Temperature:

1. Ideal Range of Temperature:

• Different temperature ranges are favorable for different plants. With hydroponic systems, farmers may adjust the temperature precisely for optimal yield.

• A wide variety of crops do best in temperatures ranging from 65 to 75°F (18 to 24°C).

2. Temperature in the Root Zone:

• To ensure that nutrients are absorbed, the root zone must be kept at the proper temperature. To do this, hydroponic systems frequently employ methods like heating or freezing nutrient solutions.

3. Differences in Day and Nighttime Temperatures:

• Plant growth and health are enhanced when natural temperature variations between day and night are replicated.

Controlling Humidity:

1. Control of Relative Humidity (RH):

• Transpiration rates and nutrient absorption are impacted by RH levels. RH should be kept between 50 and 70% typically.

• Mold and fungal problems can be caused by high humidity, which highlights the importance of adequate ventilation.

Movement Of Air:

1. The Significance of Air Movement

• To minimize stagnant air pockets around plants and lower the risk of illness, air circulation is essential.

• Fans and ventilation systems aid in the uniform distribution of heat, preserving steady temperatures throughout the hydroponic system.

2. Distribution of CO_2:

• Proper air circulation guarantees uniform CO_2 distribution, which is essential for photosynthesis. CO_2 is necessary for plant growth, and increasing its availability maximizes yields.

CO2 Improvement:

1. Levels of CO2 in Hydroponics:

• An essential element of photosynthesis is carbon dioxide. While 400 ppm is the average amount found in ambient air, hydroponic systems frequently benefit from greater levels, ideally between 800 and 1,500 ppm.

2. Methods for Injecting CO2:

In a hydroponic setting, carbon dioxide can be added by means of tanks, generators, or spontaneous fermentation.

• Monitoring devices, such as CO2 sensors, assist in preserving ideal levels and preventing over-enrichment.

3. Combining Light Cycles with Integration:

• When CO2 enrichment is timed to coincide with light cycles, it works best. Plants are guaranteed to be able to maximize photosynthesis when high CO_2 is provided throughout the light time.

In conclusion, effective hydroponic farming relies heavily on environmental management. An atmosphere where plants can thrive is created by precisely controlling temperature, humidity, air movement, and CO_2 enrichment. This leads to increased yields, faster growth, and generally healthier crops. The effectiveness and sustainability of hydroponic systems are enhanced by routine monitoring and modifications made in response to plant requirements.

CHAPTER 9

Management Of Diseases And Pests

Like any growing technique, hydroponics provides a regulated and effective environment for plant growth, but it is not impervious to problems with pests and diseases. To ensure optimal plant development and maintain a healthy hydroponic system, effective management tactics are essential. We will look at common hydroponic pests, disease prevention and control techniques, and organic pest management ideas in this tutorial.

Typical Hydroponic Pests:

1. Aphids:

• Identification: Plant sap is the food source for these tiny, soft-bodied insects.

• Management: Use insecticidal soaps or introduce natural predators, such as ladybugs.

2. Spider mites:

• Identification: Stippling on leaves is caused by tiny arachnids.

• Maintenance: Keep humidity levels high, employ predatory mites, and give your plants regular cleanings and inspections.

3. White-winged insects:

• Identification: On the underside of leaves, little, white insects can be found.

• Control: Use sticky yellow traps, let nature's foes loose, and think about using insecticidal oils.

4. Thrips

• Identification: Tiny, slender insects that consume the tissues of plants.

• Control: Apply insecticidal soap, neem oil, or beneficial insects.

5. fungus gnats

• Identification: Tiny insects that resemble mosquitoes and deposit their eggs in damp soil.

• Management: Use yellow sticky traps, let the growing medium's top layer dry out, and take predatory nematodes into consideration.

Prevention And Control Of Diseases:

1. Sterile conditions for growing:

• To stop the spread of illness, keep the hydroponic system clean and sterile.

2. Managing Nutrients Properly:

• To improve plant health and lessen susceptibility to illnesses, make sure the nutrient solution is balanced.

3. Control of Humidity and Temperature:

• Preserve ideal environmental conditions to impede the growth of bacterial and fungal illnesses.

4. Frequent Exams:

• Keep an eye out for disease symptoms in plants, such as wilting, discoloration, or strange growth patterns.

5. Vaccination Procedures:

Before adding additional plants to the hydroponic system, isolate them to stop the spread of any potential diseases.

Management Of Organic Pests:

1. Biological Regulators:

• To reduce pest populations, introduce beneficial insects like parasitic wasps, predatory mites, and ladybugs.

2. Neem Oil:

• Neem oil disrupts the life cycle of illnesses and pests by acting as a natural fungicide and insecticide.

3. Soaps that kill insects:

• They can be sprayed on plants to combat soft-bodied pests because they are made of naturally occurring fatty acids.

4. Advantageous Microbes:

• Employ fungus and helpful bacteria to improve plant health and inhibit the spread of dangerous infections.

5. Planting companion plants:

• One way to help manage pests is to plant companion crops that either attract helpful insects or repel pests.

In summary, effective hydroponic farming requires a proactive, integrated strategy for pest and disease management. Hydroponic plants will be healthier and more productive overall if regular monitoring, a clean environment, and a mix of organic and preventive management methods are used.

CHAPTER 10

Crop Rotation And Harvesting In Hydroponic Systems

1. Signs Of Ready For Harvest: To maximize productivity and quality in hydroponic systems, timing the harvest is essential. Depending on the particular crop, readiness signs might differ, but some common ones are as follows:

• Maturity Stage: Harvesting at the right time of year guarantees the best possible flavor, texture, and nutritional value because plants grow at varying rates.

• Color Changes: As fruits and vegetables ripen, many of them experience color changes. When a crop is ready for harvest, it can be determined by keeping an eye on color changes.

• Texture and Firmness: A crop's texture and firmness might reveal whether it is ready. For instance, when mature, certain fruits should feel a little bit squishy to the touch.

• Blooming and Fruiting Patterns: Predicting the harvest period is aided by knowledge of the blooming and fruiting patterns of a given crop.

• Trichomes formation: For some herbs or medicinal plants, the best time to harvest is indicated by the formation of trichomes, which are tiny structures that resemble hairs.

2. Harvesting Methods: Since hydroponic systems provide a regulated environment, harvesting can be done with more accuracy. Here are a few methods:

• Clean Tools: To stop contamination and the spread of infections, make sure that harvesting tools are sterilized and clean.

• Selective Harvesting: Only fully developed crops should be chosen for harvest. This permits other plants to develop further until they are fully grown.

• Cutting Techniques: To make precise cuts without harming nearby crops or the surviving plant, use clean, sharp equipment.

• Continuous Harvesting: By removing exterior leaves from some crops (such as leafy greens) while keeping the center growth point unharmed, continuous harvesting can be achieved for continuous production.

• Careful Handling: Take extra care while handling harvested crops to prevent bruising or damage, particularly to fragile fruits and herbs.

3. Crop Rotation In Hydroponics:

Crop rotation is the process of gradually switching up the kinds of crops cultivated in a specific area. Crop rotation is a technique used in traditional soil-based farming to maintain soil health, however, hydroponics can gain from it in several ways:

• Nutrient management: The amount of nutrients needed by various crops varies. In hydroponics, crop rotation aids in more efficient nutrient management by halting the loss of particular components in the nutrient solution.

• Disease Prevention: Even in hydroponic systems, where pathogens may still be present, crop rotation lowers the risk of soil-borne illnesses. Modifying the crops upsets possible diseases' life cycle.

• Optimal Resource Utilization: Hydroponic systems can make better use of the nutrient solution by cycling crops with varying nutrient needs, reducing the buildup of surplus nutrients that can be harmful to plants.

• Better Plant Health: By reducing stress on individual plants and halting the spread of pests and illnesses exclusive to a given crop, crop rotation helps preserve the general health of plants.

• Enhanced System Productivity: The hydroponic system's overall productivity can be raised by switching between crops with various growth patterns in order to maximize space use.

In conclusion, good hydroponic farming promotes healthy plants and maximizes yields through the use of crop rotation plans, suitable harvesting procedures, and an awareness of the indicators that a plant is ready for harvest.

CHAPTER 11

Promoting Your Hydroponic Vegetables

The success of your hydroponics business depends on how well you market your produce grown hydroponically. Understanding your target market, developing a distinctive brand identity, selecting the best sales and distribution channels, and putting Internet marketing tactics into practice are all necessary for effective marketing. Below is an explanation of each idea:

1. Choosing The Market You Want To Enter:

a. Investigation and Evaluation:

• Find out about the preferences of potential clients by conducting market research.

• To better understand your target audience, examine psychographics, buying patterns, and demographic data.

b. Adapt Products to the Needs of the Market:

• Tailor your hydroponic vegetable selections to your target market's recognized needs and preferences.

• Take into account variables such as dietary limitations, lifestyle decisions, and taste preferences.

c. Create a USP or unique selling proposition:

• Set your hydroponic produce apart from the competition by emphasizing specific features like organic, regionally grown, or unusual types.

• Make it obvious why selecting your produce is superior to more conventional options.

2. Establishing A Personal Brand:

a. Packaging and Logo:

• Create a logo that captures the spirit of your hydroponics company and is visually appealing.

• To increase brand visibility and appeal, spend money on aesthetically pleasing and sustainable packaging.

b. Brand Communication:

• Create a brand narrative and message that appeals to your target audience.

• Highlight the advantages of produce grown hydroponically for the environment, human health, or ethical reasons.

c. Uniform Branding Throughout All Platforms:

• Make sure that the branding is consistent across all channels, such as print advertising, web content, and physical packaging.

• To increase brand recognition, and uphold a consistent brand image.

3. Channels For Sales And Distribution:

a. Choosing Distributor Partners:

• Distribution channels should be chosen with your target market in mind. Grocery stores, farmers' markets, and specialty shops may be examples of this.

• Build connections with chefs and eateries in the area to get your produce featured.

b. Sales directly to consumers:

• Take into account going direct-to-consumer via an internet platform, farm stand, or subscription service.

• Organize workshops or events to interact with the neighborhood and develop a clientele.

4. Internet Marketing Techniques:
a. Social Media Awareness:

• Make use of social media channels to interact with your audience, share farming techniques, and display your hydroponic produce.

• Make sure your products are aesthetically appealing by using videos and images to showcase their high caliber and freshness.

b. Internet and Online Sales:

• Create an easy-to-use website with details about your hydroponics business, products, and ordering procedures.

• Put e-commerce features into place to make online ordering and delivery easier.

c. Marketing with Content:

• Produce educational content about sustainable farming, healthy eating, and hydroponics to establish your brand as an authority in the field.

• Use podcasts, blogs, and newsletters to provide your audience with insightful content.

d. Internet Promotion:

To reach a larger audience, spend money on social media or Google Ads, two platforms that offer targeted online advertising.

• Monitor and assess the effectiveness of your online marketing initiatives to improve performance.

You can effectively promote your hydroponic produce, establish a strong brand presence, and cultivate a devoted customer base by strategically putting these marketing concepts into practice. To guarantee long-term success, evaluate and modify your marketing tactics frequently in response to consumer input and industry developments.

Summary

To sum up, learning about hydroponics has opened up a world of information and opportunities for growing plants without soil. The experience with this cutting-edge farming technique has given me new perspectives on effective and sustainable crop-growing techniques. It's critical to summarize the most important lessons learned, provide motivation for future achievement, and provide closing remarks and acknowledgments as we draw to a close our investigation.

Summary Of Important Takeaways:

As we've investigated hydroponics, we've discovered that it has a number of benefits over conventional soil-based farming. Exact control over nutrient levels, water use, and environmental factors is made possible by hydroponic systems. Higher yields from crops, quicker rates of growth, and the capacity to cultivate plants in regions with subpar soil are all results of this enhanced control.

We have also studied a variety of hydroponic systems, each with pros and cons of their own, including aeroponics, deep water culture (DWC), and nutrient film technique (NFT). One of the most important things to know about optimizing plant growth in hydroponic solutions is the significance of pH and nutrient balance.

Furthermore, the idea of vertical farming—growing crops in layers that are stacked—has been investigated as a way to maximize space efficiency and solve the problems associated with urban agriculture. The accuracy and effectiveness of hydroponic systems are further improved by the incorporation of technology, such as automation and sensors.

Motivation For Upcoming Achievements:

Hydroponics has the potential to completely transform agriculture in the future. This creative farming method has the potential to lessen the negative environmental effects of conventional agriculture, reduce water consumption, and address the issue of food security. Hydroponics is versatile enough to be used in many different contexts, from large-scale commercial operations to small-scale home gardening.

Promoting hydroponics among people and communities can help advance sustainable agricultural methods. To realize the full potential of hydroponic techniques, research and education investments are essential. With the world's population expanding and climate changing, hydroponics provides a way to produce food that is more resilient and uses fewer resources.

Closing Remarks And Recognitions:

As we progress through the hydroponics process, it is crucial to consider the combined efforts and contributions of scientists, innovators, and enthusiasts. Hydroponic technology and practices have advanced due to the combined efforts of multiple disciplines, such as biology, engineering, and agriculture.

We are especially grateful to the visionaries and trailblazers who made hydroponics a viable and sustainable agricultural method. A new generation of farmers and gardeners has been motivated to investigate alternative cultivation techniques by their commitment and enthusiasm.

Let's embrace hydroponics' potential to revolutionize food cultivation as we go forward. We can build a future in which hydroponics is essential to maintaining food security, environmental preservation, and the welfare of communities everywhere by promoting an innovative and sustainable culture.

THE END